existence preserves existence

That is all there is to it ... but how do you write this down and why ?

Let's get started ...

Existence preserves existence

Three thought experiments :

> I have omitted some scenarios ... in the thought experiments ... randomness exists and in some experiments ... randomness would yield more results ... I know ... this is to give an example ...

A

Get 10 people to pick a green vase from a room where are there are ten differently colored vases ...

One of the people is colorblind ...

How many green vases will you get ?

9 ... except sometimes by chance you will get 10 or 8 (or any other chance event)

And this teaches us that there is something in the Universe that is "green" ... this ATTRIBUTE is common, but we have no idea if the green you see is the green I see ... therefore there is something called "green" ...

Tara Vanhonacker .... Ff 2017

B

Get 10 people to pick a green vase from a room where there are ten differently colored vases ... One of the people is blind ....

How many green vases will you get ?

9 ... except sometimes by chance you will get 10 or 8

And this teaches us that there is something in the Universe that is "chance" ... this ATTRIBUTE is common, but we have no idea why the chance happens therefore there is something called "chance" ...

C

Get 10 people to pick a vase from a room with three plates and a vase on display ... One of the people is blind ...

How many vases will you get ?

10 ... except sometimes by chance you will get 9

And this teaches us that there is something in the Universe that is "vase" ... this ATTRIBUTE is common, but we have no idea if the vase you see or feel is the vase I see or feel ... therefore there is something called "vase" ...

Tara Vanhonacker …. Ff …. 2017

This is part one of the logic ...

OBJECTS + ATTRIBUTES

- vase
- green
- chance

Remark :
we get … "green" and "vase" which are patterns that repeat themselves and have common attributes, they however are also attributes of the object "object" …

repetition is what occurs normally and chance is what we perceive as accidental random abnormal events

A second thought experiment :

> In a totally random universe you could get whatever result you wanted ... statisticians will agree with me, that I used a most-likely scenario....

Imagine a room with 10 food utensils and within those ten utensils there is one vase ... you get 10 aliens who have no sense of sight and you tell them to get the vase ... You will get 0 ... maybe by accident 1 ... and why ... because a species that depends on other senses than sight, will have trouble identifying an object that we can see ...

Does the object "vase" exist ... yes it does ... but the way to describe that object will be different for this alien species ...

And now it comes and this is a fun part ...

"sight sense" is also an attribute of a higher object which we could call ... "perception" ...

OBJECTS + ATTRIBUTES

- vase
- green
- chance
- perception

What do we know NOW :
- there is something called "green"
- there is something called "vase"
- there is something called "perception"
- there is something called "chance"

Tara Vanhonacker .... Ff 2017

What does this mean to our view of the Universe ?

1) objects exist, outside of our mind, there is a universe of "things"

2) objects have attributes and at the same time attributes can be objects

3) "perception" is tricky, we will have different ways to describe the Universe depending on the senses we use and the limitations of these senses

4) there is randomness, or chance

Now comes the question, what is an object ?

What is "perception" ?
What is an attribute ?

I will expand later upon the reasoning that Time is an attribute.

$E = m*c^2$ ...

C^2 is not a velocity, it is merely a constant.

This has led me to this :

$En/Ms = Ff$ ... or C^2 (the latter one being confusing)

(C^2 refers too many times to C in our minds, which is the velocity of light ... hence the confusion when you meditate on the Universe, it's limiting our perception and logic ... so I call it Ff ... the link with C is lost then and that leaves us to think more clearly.... $E = M*Ff$)

Explained : **Every object that exists is a manifestation of a ratio of Energy and Mass ... Ff ...**
I have omitted Time in this equation because I believe Time is an attribute and not a constant ...

1 VASE = En/Ms = 5472358/25415871 (This is just an example)

A vase is an object but it is not a definite object ... it is actually ... and we know this ... a set of molecules that work together to construct a physical object that has the attributes of a vase ... and that is an Ff ... Ff is nothing more than a program or a set of instruction codes that are manifested in the laws of physics, chemistry ... and so on ...

Tara Vanhonacker …. Ff …. 2017

Let's go to particles …

En / Ms = Ff …

A photon is still En/Ms … we have trouble detecting the mass in this situation, but it is there, minutely … Ff is for instance the rest point of a photon as described in (quantum) physics …

Again I omit TIME …

En / Ms is the object and Ff is the attribute …

Object A (*photon*) **= En/Ms = Ff**

Now imagine we have two objects (like two photons, or planets or vases)

En / Ms = Object1 = Ff

En^2 / Ms^2 = Object2 = Ff

Object2 - Object1 = Ff … a set of instructions

Which means that instead of looking at results and begin-situations, we should start looking at the instruction set …

I want to point out that is difficult to write down because of the limitations of my keyboard.

Why is there an instruction set ?

That is fun ...

Ff is just an expression of existence ... look at this weird equation :

En = 0 , Ms = 10 ... Ff = 0 ...

En (0) / Ms (10) = Ff (0)

The instruction set is empty ... which means that the En/Ms does not exist ... is not in existence ...

En = 10 , Ms = 0 ... Ff = nan or !error ... is not in existence ...

So existence NEEDS a set of instructions or does not exist ...

The constant of the Universe is Ff ... Ff ... flow ... existence ... life ...

About time ...

Time is an attribute ...

Math :

En = 10

Ms = 0

Ff = nan (error)

En = 0

Ms = 10

Ff = 0

So we can write this down :

(this is pseudocode)

if (En = 0)

{

if (Ms = 0)

{

Ff != !error … <non - existent>

}

}

This is not the full instruction set … only a snippet.

Tara Vanhonacker .... Ff 2017

What is an object ?

An object is the representation of an instruction set, which means that we could write this down :

Object A = En / Ms = Ff

An object always has attributes ...

- temperature
- time
- weight
- gravitation
- length
- width
- location
- color
- age
- mass
- energy
- ....

(We can go on for a long time)

However we are used to Energy and Mass as they are the most universal ways to measure or detect objects and why, because our sense, lead to easy recognition of both) ...

An object always has attributes ...

- temperature
- time
- weight
- gravitation

- length
- width
- location
- color
- age
- **mass**
- **energy**

We could have defined an object by its weight and temperature … why not … yet in modern physics we have chosen to use Ma and En … as the most important attributes to define an object.

In the case of an alien species with no sense of sight … we might conjecture that **mass** and **temperature** could be used … or **sound** … but we are not that alien species.

Object A = En/Ms = Ff

And this is bound to the rules that En and / or Ms can never be 0.

If En/Ms = 0 or nan => Ff => 0 or nan => Object A => 0 or nan

Which means it does not exist …

The big bang according to Professor Hawking

Professor Hawking states in an extensive interview that there was first "nothing" and then there was the "universe" … more or less like a light switch … and he is correct …

When there is nothing : $En = 0$, $Ms = 0$ then Ff = nan or non-existent or 0 and equally non-existent ….

Upon the Big Bang the Universe got En = something and Ms = something which in return means there is an Ff. Again Time is an attribute …

Universe = En/Ms = Ff

At the same time we can observe this for a single Photon :

Photon = En/Ms = Ff

The universe and a photon are different, but have the same attributes … from the infinitely small .. to the vast and seemingly random and chaotic large …

Science and language

When we write down En/Ms = Ff, the Ff contains a value or it contains an instruction set and this instruction set can be described by several languages :

- natural speech

- Physics

- Geometry

- Math

- Chemistry

- Biology

- Mechanics

- Quantum

and many more …

So Object A = En/Ms = Ff in which the Ff could be

a chemistry formula and at the same time a physics formula …

We will see that we will need to find common ground and make these separate languages converge into a scientific lingua franca or we need to create a new language … I would personally prefer to use Math … but it is not really up to me …

I also want to point out this from pseudocode :

Tara Vanhonacker .... Ff 2017

#include <physics.h>

int main()
Float Em, Ms, Object A;
{

En / Ms = Object A;

En /Ms = {Physics code};

Return 0;

}

Between {Physics Code}
you could find some formula, measurement, quantification ... in the language of physics.

#include <chemistry.h>

int main()
Float Em, Ms, Object A;
{

En / Ms = Object A;

En /Ms = {Chemistry code};

Return 0;

}

Between {Chemistry Code} you could find some formula, measurement, quantification ... in the language of chemistry

Tara Vanhonacker …. Ff …. 2017

Here is the beauty :

#include <chemistry.h>

#include <physics.h>

int main()

Float Em, Ms, Object A;

{

En / Ms = Object A;

En /Ms = {Chemistry code};

..OR..

En / Ms = {Physics code};

/* {Chemistry code = Physics Code} */

Return 0;

}

Because we are talking about ONE object, A, and we have physics as well as chemistry explanations for Object A, we can carefully state that the code of physics and chemistry are

the same as there is only one Object A.

Of course in speech and writing it will sound different and it has an altogether different notation, but they are the same …

So we could say this :

En^1/Ms^1 = Object A

This means we have an Object A, that is a certain ratio of En by Ms. The ^{1} is not an exponent. I have at present no other way to write down a clear notation … on a computer ….

En^2/Ms^2 = Object B

This means we have an Object A, that is a certain ratio of En by Ms. The^2 is not an exponent. I have at present no other way to write down a clear notation … on a computer ….

The difference between Object A and Object B is the same as :
$En^1/Ms^1 - En^2/Ms^2$ also …. Ff^1-Ff^2
or going back ti our pseudocode :

$\{Physics\ code\}^1 - \{Physics\ code\}^2$

or … $\{Chemistry\ code\}^1 - \{Chemistry\ code\}^2$

or … $\{Physics\ code\}^1 - \{Chemistry\ code\}^2$

or vice versa …

Which means that we have two methods of describing the difference between Object A and Object B … or the evolution or change from Object A into Object B … or many more ...

This looks simple and obvious but we need a *"superlanguage"* that surpasses present scientific and separate languages (Math ?).

I think it should be done by computer and computer scientists should assist in creating this "database of Ff" … I refer to technologies such as the distributed SETI-project… we could do the same for science and find patterns that we have not yet been able to see …

End of Part ONE …

Part two comes later this week and has a few surprises …

Tara Vanhonacker .... Ff 2017

Why does this process of photons binding into molecules happen ?
What else can we do with photons ?

https://news.harvard.edu/gazette/story/2013/09/seeing-light-in-a-new-way/

Scientists from Harvard University and the Massachusetts Institute of Technology (MIT) are challenging the conventional wisdom about light, and they didn't need to go to a galaxy far, far away to do it.

Working with colleagues at the Harvard-MIT Center for Ultracold Atoms, a group led by Harvard Professor of Physics Mikhail Lukin and MIT Professor of Physics Vladan Vuletic managed to coax photons into binding together to form molecules — a state of matter that until recently had been purely theoretical. The work is described in a Sept. 25 paper in Nature.

The discovery, Lukin said, runs contrary to decades of accepted wisdom about the nature of light. Photons have long been described as massless particles that don't interact with each other. Shine two laser beams at each other, he said, and they simply pass through one another.

Photonic molecules, however, behave less like traditional lasers and more like something you might find in science fiction: the light saber.

"Most of the properties of light we know about originate from the fact that photons are massless, and that they do not interact with each other," Lukin said. "What we have done is create a special type of medium in which photons interact with each other so strongly that they begin to act as though they have mass, and they bind together to form molecules. This type of photonic bound state has been discussed theoretically for quite a while, but until now it hadn't been observed.

"It's not an inapt analogy to compare this to light sabers," Lukin said. "When these photons interact with each other, they're pushing against and deflecting each other. The physics of what's happening in these molecules is similar to what we see in the movies."

To get the normally massless photons to bind to each other, Lukin and his colleagues, including Harvard postdoctoral fellow Ofer Firstenberg, former Harvard doctoral student Alexey Gorshkov, and MIT graduate students Thibault Peyronel and Qiu Liang, couldn't rely on something like the Force. They instead turned to a set of extreme conditions.

Researchers began by pumping rubidium atoms into a vacuum chamber, then used lasers to cool the cloud of atoms to just a few degrees above absolute zero. Using extremely weak laser pulses, they fired single photons into the cloud of atoms.

As the photons enter the cloud, Lukin said, their energy excites atoms along its path, causing the photons to slow dramatically. As the photons move through the cloud, that energy is handed off from atom to atom, and eventually exits the cloud with the photon.

"When the photon exits the medium, its identity is preserved," Lukin said. "It's the same effect we see with refraction of light in a water glass. The light enters the water, it hands off part of its energy to the medium, and inside it exists as light and matter coupled together. But when it exits, it's still light. The process that takes place is the same. It's just a bit more extreme. The light is slowed considerably, and a lot more energy is given away than during refraction."

When Lukin and his colleagues fired two photons into the cloud, they were surprised to see them exit as a single molecule.

The reason they form the never-before-seen molecules? It's an effect called a Rydberg blockade, Lukin said, which means that when an atom is excited, nearby atoms cannot be excited to the same degree. In practice, the effect means that as two photons enter the atomic cloud, the first excites an atom, but it must move forward before the second photon can excite nearby atoms.

The result, he said, is that the two photons push and pull each other through the cloud as their energy is handed off from one atom to the next.

"It's a photonic interaction that's mediated by the atomic interaction," Lukin said. "That makes these two photons behave like a molecule, and when they exit the medium they're much more likely to do so together than as single photons."

While the effect is unusual, it has some practical applications.

"We do this for fun, and because we're pushing the frontiers of science," Lukin said. "But it feeds into the bigger picture of what we're doing because photons remain the best possible means to carry quantum information. The handicap, though, has been that photons don't interact with each other."

To build a quantum computer, he said, researchers need to build a system that can preserve quantum information and process it using quantum logic operations. The challenge, however, is that quantum logic requires interactions between individual quanta so that quantum systems can be switched to perform information processing.

"What we demonstrate with this process allows us to do that," Lukin said. "Before we make a useful, practical quantum switch or photonic logic gate, we have to improve the performance. So it's still at the proof-of-concept level, but this is an important step. The physical principles we've established here are important."

The system could even be useful in classical computing, Lukin said, considering the power-dissipation challenges that chip-makers face. A number of companies, including IBM, have worked to develop systems that rely on optical routers that convert light signals into electrical signals, but those systems face their own hurdles.

Lukin also suggested that the system might one day even be used to create complex, 3-D structures, such as crystals, wholly out of light.

"What it will be useful for we don't know yet. But it's a new state of matter, so we are hopeful that new applications may emerge as we continue to investigate these photonic molecules' properties," he said.

What is chance ?

It looks like we experience physical phenomena we can not explain ... at present, so we call it "chance" ... in my opinion in later centuries we will find an explanation... does it hinder us ... no ... we can predict for a fair amount of time ... the behavior of so called chance events and results ...

www.ingramcontent.com/pod-product-compliance
Lightning Source LLC
Chambersburg PA
CBHW051836210526
45473CB00005B/1907